P9-DZZ-771

IMAGINE LIVING HERE

THIS PLACE IS
COLD

BY
VICKI COBB

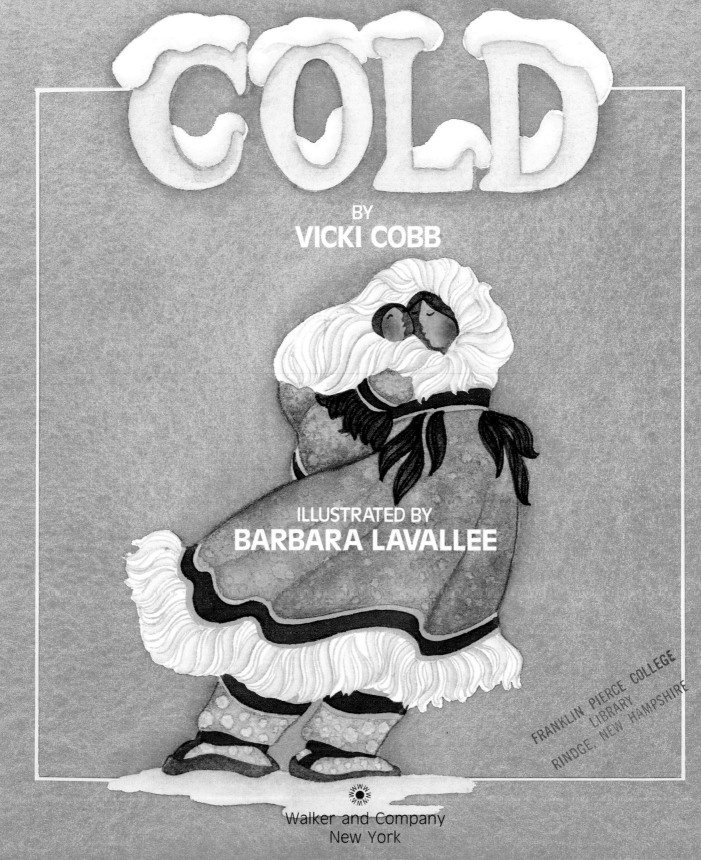

ILLUSTRATED BY
BARBARA LAVALLEE

FRANKLIN PIERCE COLLEGE
LIBRARY
RINDGE, NEW HAMPSHIRE

Walker and Company
New York

For Carol and Vic Hussey

The author gratefully acknowledges the assistance of the following Alaskans: Jan Westfall, Dee McKenna, Roz Goodman, Jane Behlke, and Jim and Emma Walton.

Copyright © 1989 by Vicki Cobb
Illustrations copyright © 1989 Barbara Lavallee

All rights reserved. No part of this book may be reproduced or transmitted in any form or by any means, electronic or mechanical, including photocopying, recording, or by any information storage and retrieval system, without permission in writing from the Publisher.

First published in the United States of America in 1989 by the Walker Publishing Company, Inc.

Published simultaneously in Canada by Thomas Allen & Son Canada, Limited, Markham, Ontario.

Library of Congress Cataloging-in-Publication Data

Cobb, Vicki.
This place is cold! / by Vicki Cobb ; illustrated by Barbara Lavallee.
p.   cm.—(Imagine living here)
Summary: Focuses on the land, animals, plants, and climate of Alaska, presenting it as an example of a place where it is so cold your hair can freeze and break off.
ISBN 0-8027-6852-0 ISBN 0-8027-6853-9 (lib. bdg.)
ISBN: 0-8027-7340-0 (PBK)
1. Alaska—Description and travel—1981—Juvenile literature.   2. Cold regions—Juvenile literature.   [1. Alaska—Description and travel.
2. Cold regions.] I. Lavallee, Barbara, ill.   II. Title.
III. Series: Cobb, Vicki. Imagine living here.
F910.5.C63   1989
917.98—dc19
88-25919
CIP
AC

Printed in Hong Kong

First paperback edition printed in 1990

10 9 8 7 6 5 4

Text design by Laurie McBarnette

CURR
F
910.5
.C63
1990

If you visited Alaska, a place near the top of the world, during the winter the first thing you would notice is the cold and the darkness. How cold can it get? Temperatures can drop fifty degrees below zero and even lower. The engine blocks of cars have special built-in heaters. People plug their cars into electrical outlets so that their engines won't get too cold to start. Firewood is so frozen that one tap of an ax shatters a log. Snow squeaks loudly when you step on it. Sometimes the wind is so strong you have to walk at a slant. Your breath turns instantly into tiny ice crystals that glitter in the sun. Your eyebrows and eyelashes freeze, but you don't dare rub them for fear they will break off.

In the winter, the North Pole points away from the sun, so the top of the world, called the Arctic Circle, is always in darkness. Just below the Arctic Circle, in the heart of winter, the sun comes up around lunchtime for only a few hours. Children wake up when it's dark, go to school in the dark, and come home in the dark. On some winter nights the sky is decorated by an incredible display of swirling colored curtains of light called the *aurora borealis* or northern lights.

Water everywhere is frozen. Millions of years ago, rivers of ice, called glaciers, began to form all over Alaska. Glaciers form in the valleys between mountains as snow piles up from year to year. Since it is always cold, most of the snow doesn't melt. The weight of the piled snow changes the snow into ice. The ice is packed so hard that it changes color. It is no longer clear like an ice cube. The glacial ice is now a bright blue-green. Glaciers are nature's giant bulldozers, and do they ever carve up the land!

The heavy, old ice starts moving very slowly down the mountain, carrying along rocks. It grinds up the soil, which ends up like powder, finer than sand. Glacial ice full of soil is black. It takes millions of years for glaciers to carve up the mountains. A valley that was shaped like a V becomes broad and wide at the bottom like the letter U. In some places, the land looks as if a giant melon scoop has been at work.

Some glaciers end at the sea or a lake. The end of such a glacier keeps breaking off into chunks of floating ice called icebergs. Some glaciers melt. The dirt is left behind as a huge, long, high hill. The melting glacial waters run off in many twined curved streams forming "braided" rivers.

In some places ice in the ground doesn't melt even in the summer. This kind of land is called *permafrost* because it is permanently frozen. Here and there, deep in the permafrost soil, are blocks of ice called ice lenses. If a house is built over permafrost or an ice lens, the heat from the house may cause melting. The water seeps away and the thawed ground sinks. The part of the house that rests on the thawed ground will sink too. Sometimes to straighten out a house, people use jacks like the ones you use to change tires, only stronger. You can't count on the floors staying straight or doorways staying square. So some people use curtains inside instead of doors.

In many places, the permafrost is covered with a thick layer of short, scrubby plants, like blueberry bushes. The land is called the *tundra* which comes from a Russian word meaning ``bare hill.'' The heat from the sun is trapped by the plants and doesn't pass into the soil. So the tundra plants keep the permafrost from getting warm enough to melt.

The plants that grow on the tundra must be able to survive the extreme cold and dryness. There are no large trees here because even in the summer when the ground thaws a little, there isn't enough water for them to grow. The most ancient plants are the scrubby *lichens* (LI-KENS). Lichens have survived for millions of years because they are a partnership of two plants. One plant is a fungus. A fungus cannot make its own food, but its shape allows it to catch rainwater. The other plant has no shape that you can see. It is millions of tiny green plants called *algae* (AL-gee), which live so closely with the fungus that they seem to be one plant. The algae make the food the fungus needs, while the fungus collects the water the algae need to make food. Lichens can grow on rocks since they have no roots and don't need soil. Lichens grow very, very slowly. A lichen only two inches across may be hundreds of years old.

One kind of lichen is called *reindeer moss*. This might be because it is shaped like a reindeer's antlers or maybe because reindeer eat it. The tundra is the summer grazing land of reindeer and their slightly larger cousins, the caribou. It is also the home of the largest member of the deer family, the moose. The male, called a bull, can be seven and a half feet tall at the shoulders and weigh almost a ton. His antlers alone can weigh a hundred pounds. Moose shed their antlers and grow a new set every year. Bulls use their antlers to fight each other over the females, called cows.

The musk-ox, a distant cousin of the wild goat, also eats plants of the tundra. The musk-ox looks the same as it did millions of years ago. It gets its name from the sharp, sweet "musky" smell that the males give off. When danger threatens, musk-oxen herds are noted for standing shoulder to shoulder in a circle with their young in the center. This stand protects the young from wolves and other hunters. But it is no protection from human guns and arrows. A hundred years ago the musk-oxen were hunted almost to extinction for their meat and their wool.

The musk-ox has a long, shaggy coat that sways like a skirt in the breeze. Next to the body is a thick layer of wool that is eight times warmer than sheep wool and softer than cashmere. This wool is shed in patches in the summer. Eskimo women gather this wool and comb tame musk-oxen to collect more. They knit this fine wool into very expensive hats and scarves. Laws protecting the musk-ox have helped this animal make a remarkable recovery. Today, musk-oxen are raised for their wool on ranches and some have been set free to roam the tundra again in wild herds.

The most dangerous animals in Alaska are the bears. The world's largest meat eater is the brown bear. It weighs almost a ton, a powerful sixteen hundred pounds. But no one wants to run into a hungry grizzly or polar bear either. All bears kill small animals for food, and some will attack a sleeping camper and raid a tent for food. People who travel in bear territory learn to respect bears. Some hikers use a walking stick with bells that ring with each step so they won't surprise the bears. These people never eat in the same place they camp because bears might be attracted to the smell of food and surprise them while they're sleeping. In addition to meat and fish, bears have a sweet tooth and love berries and honey. They eat well through the summer and fall to put on enough weight so they can sleep through the coldest winter months in their snug dens.

White polar bears have fur that is different from all other animals. Close to the body is a thick layer of wooly fur for staying warm. Long *guard* hairs stick out of the fur. The guard hairs look as if they are made of plastic. They are hollow like straws and always stick out. So when the polar bear swims in the icy water while hunting for seals or fish, the guard hairs keep the fur from getting matted down. Water is easily shaken off the polar bear's stiff outer coat.

A polar bear's white coat is protective camouflage in the ice and snow. But white reflects sunlight. It is a color for hot climates, not cold. Black is the color that absorbs heat, so nature gave polar bears black skin. The guard hairs trap heat from the sun and conduct it to the skin, which absorbs it. Under the skin is a layer of fat that also keeps the animal warm.

Polar bears have thick fur between the pads on the bottoms of their feet, to keep their feet from freezing as they walk over the ice. Like other bears, polar bears are fierce and dangerous. Killing a polar bear was a tradition for Eskimo boys on their way to becoming men.

Polar bears are hunters and fishers. They hunt sea mammals like small whales and walruses. But their favorite food is the seal. It takes a lot of skill for a polar bear to catch a seal. There's no way the bear can outswim a seal to catch it. Polar bears attack seals when they are resting on the ice or when they come up for air at the edge of the ice.

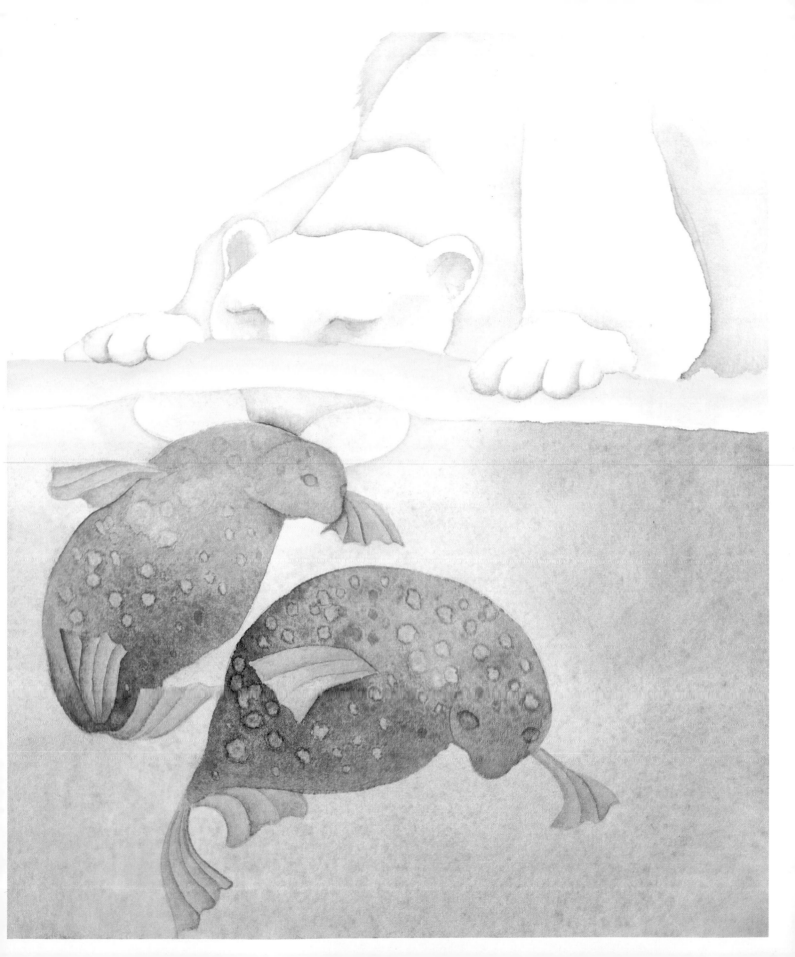

But polar bears are not the worst enemy of sea mammals. People have hunted them for centuries. Seals have been hunted for their shiny, thick fur. Walruses have ivory tusks that they use to dig up clams and mollusks on the bottom of the sea. Unluckily for the walrus, the ivory can be carved into tools, jewelry, and small statues. Today, to protect the walrus, only Eskimos are allowed to hunt them and carve the ivory.

All sea mammals have a layer of blubber under their skin to keep them warm. Eskimos hunted sea mammals for their blubber as well as for their meat. Eating animal fat helped the Eskimos make a layer of fat under their own skins to stay warm. Whales, the largest mammals of all, have the most blubber. And it was for blubber that they once were hunted. The blubber of whales can be cooked down into a fine oil that people once used in lamps to light their homes.

Instead of teeth, one kind of whale has a mouth full of long, flexible plates with fringes trailing behind . These plates are called baleen and the whale is called a *baleen whale*. It eats by swimming along with its mouth open, taking in water containing millions of tiny plants and animals that float in the ocean. Some baleen whales eat up to four and a half tons of this stuff a day! So a baleen whale doesn't need teeth. It needs a strainer to trap its food and that's what the baleen is.

Before plastic was invented, baleen was used to make umbrella ribs and eyeglass frames and the stiff supports of ladies underwear. Baleen can also be peeled into long thin strips. Eskimos weave these strips into beautiful baleen baskets.

Fishing is more than just a sport in the far north. For certain animals and people it is a skill that is necessary for survival. There are many different kinds of fish in Alaskan seas and rivers. But the most popular and amazing is the *salmon*.

A full-grown red salmon weighs about ten pounds and lives in the ocean. But it didn't begin its life in the ocean. Salmon eggs hatch in the fresh waters of streams, known as "spawning grounds." Baby salmon live in their spawning grounds until they are two years old. Then they travel downstream. The streams get larger and become rivers until finally they join the ocean where the salmon spend their adult life. When a red salmon is about four years old it is ready to lay eggs. So it begins its return journey back to the exact place where it hatched. This means traveling upstream for hundreds of miles, swimming against the current, leaping over waterfalls, in an exhausting trip to its spawning grounds. The females lay their eggs, the males fertilize them, and then the adults die.

Bears and fishermen know that the best way to fish for salmon is to catch them as they make their last journey upriver. Bears eat most of what they catch. But people can be greedy. Today there are laws that control who may fish for salmon and how many fish they can take. The laws protect the fish to make sure there will be plenty of salmon in the future.

The first people to settle Alaska were Indians and Eskimos. They were expert hunters and fishermen. They traveled over the snow with sleds drawn by teams of dogs. They made boats and clothing from animal skins. They told stories, made up songs, and created art that showed how they lived and how they felt about their lives.

Why would other people come to this harsh land where they would have to put up with the cold, the snowstorms, the dangerous animals, and the aloneness? In the beginning, some people came for adventure, others for riches.

The first people to come from the outside world were whalers and trappers. Whalers wanted oil and baleen. Trappers went after the animals. The thick, shiny furs of seals, otters, beavers, and arctic foxes and wolves brought high prices in the world of fashion. The furs that were perfect for keeping arctic animals warm became coats worn by stylish women in climates that were not nearly as cold.

Around the beginning of the twentieth century, gold was discovered in the middle of Alaska and on the beaches of the Bering Sea. Thousands of hopeful gold miners rushed to Alaska. The people who sold them supplies were not far behind. Besides gold, a giant copper mine and the millions of salmon promised wealth. Some of the people who came to get rich did, although most didn't. But many of them stayed. They married and had children. They made new lives in small villages.

All these people needed ways to get around. A railroad was built to connect the main cities of Anchorage and Fairbanks. But there were many people who lived in the "bush" in towns like Unalakleet and Shaktoolik. Places you can reach only by dogsled, snowmobile, boat (after the ice melts), or airplane are called the bush. Villages in the bush created an opportunity for a new kind of work, namely the bush pilot. Bush pilots are skilled in taking off and landing on dirt runways or on runways cleared on frozen lakes. They bring food, medicine, mail, and contact with the outside world.

It's no big deal to have a pilot's license in Alaska. One person in every forty-five has one. There are more private planes there than in any other part of the world. Alaskans think of their planes as you might think of a family car. People use planes to go to vacation cabins, to visit friends in bush villages, and to get to a major airport to catch a jet to the rest of the world.

In 1969 a "supergiant" oilfield was discovered in the Arctic Circle. Oil has been called "black gold" because the modern world depends on oil for so many things. The problem with the Arctic oil was not how to pump it out of the frozen tundra, but how to get it to the rest of the world. There were no highways or railroads to this remote part of the world. And the nearby sea was frozen for nine months of the year. The only solution was to build a pipeline eight hundred miles long to pump the oil to an ice-free port where super tankers could load the oil.

Eight oil companies got together to build the pipeline. It was the world's most expensive and challenging project paid for by private money rather than government taxes. The work would be hard and uncomfortable. Workers would have to spend long hours in the cold, lonely wilderness. But it would pay well, very well. So thousands of young people came from all over America and the world. A worker would wear three layers of clothing under down overalls and parka, two layers of gloves, three pairs of socks, a hat under a hardhat, and a face scarf. It took ten years and more than seventy thousand men and women to build a pipeline that is one of the wonders of modern engineering. Building the pipeline was a modern "gold rush" that brought many new settlers to this northern land.

Some of the settlers were farmers. Summer in the far north has more than enough sunlight to grow certain crops like cabbage and potatoes. The summer is just the opposite of the winter. The North Pole points toward the sun and the Arctic Circle is always in sunlight. Inside the Arctic Circle, the sun doesn't set from March 21 to September 21. It just keeps making circles around the sky. Just below the Arctic Circle the sun shines almost all night long. Imagine standing with your back to the rising sun on a June day. The sun rises over your left shoulder, makes almost a complete circle around the sky and sets behind your right shoulder twenty-two hours later. The two hours it dips below the horizon do not make the night very dark. This was enough daylight for a farmer to once grow a cabbage that weighed eighty pounds!

All this daylight changes the way people act. No one wants to go to sleep. You need dark window shades to make it feel like bedtime. Children feel like playing all night long. You could plan a baseball game for eleven o'clock in the evening. In spite of all the sunlight, the summers near the Arctic Circle are not very warm, although sunny summer days feel warm to the people used to living there. Children in Alaska begin wearing shorts to school when the temperature is about fifty degrees. In a few places, like Fairbanks, summer temperatures may go as high as eighty-five or ninety.

Why do people want to live in a land where winters seem endless and most of the land is wilderness? For some it is the beauty of the wilderness itself. For some it is the challenge of settling a frontier, being the first to make a civilization.

For some, life here is a test of themselves. They take pride in making a life of joy and purpose in a harsh land for there is certainly plenty of work to be done. This is a place of opportunity for doers. Life here feeds the imagination. Can you imagine living here?